Patchouli Herb

Commercial Growing Practices and Economic Importance

Roby Jose Ciju

i

CONTENTS

PATCHOULI

Introduction

Scientific name of Patchouli is *Pogostemon patchouli* (Syn. *Pogostemon cablin*). Patchouli belongs to mint family i.e. Labiatae or Lamiaceae. Patchouli is a tropical, perennial aromatic herb which is mainly grown for its aromatic leaves. Patchouli oil, a highly fragrant essential oil, is extracted from its dried herbage. Patchouli is also known as Stink Weed or Pucha Pot (Putcha-Pat). There are several species of patchouli under cultivation such as *Pogostemon patchouli, Pogostemon commosum, Pogostemon hortensis, Pogostemon heyneasus* and *Pogostemon plectranthoides*. However, *Pogostemon patchouli* yields the highest quality patchouli oil and is, therefore, widely cultivated for commercial purposes.

Taxonomy

Kingdom	Plantae/ Angiosperms
Order	Lamiales
Family	Lamiaceae / Labiatae
Genus	Pogostemon
Species	P. patchouli/ P. cablin

Botanical Description

Plant
Patchouli is a short-day plant and is shade-loving. It grows well under partial shade conditions. The plant is a perennial, small bushy herb with branched stems. The plant reaches up to a height of about one meter upon full growth.

It is tropical in growth habit. It is commercially cultivated for its highly fragrant leaves which yield famous patchouli oil under steam distillation process.

Stems
As typical in case of all mint herbs, patchouli has quadrangular erect stems which are covered with hairs and plump nodes.

Leaves
Leaves are pale green in colour and ovate to ovate-oblong in shape. It is glossy in appearance, soft and oppositely arranged. Leaves are with lobed, serrate margins. Hairs are present on both sides. They are highly aromatic due to the presence of essential oil.

Flowers
Plants bear small, pale pink-white flowers both on its axillary and terminal spikes. Flowering season is between February and March.

Origin and Distribution

Patchouli is believed to be originated in Indo-Malayan tropical regions of Asia.

Patchouli grows well in all warm to tropical climates and is, therefore, extensively cultivated in tropical climates of Caribbean countries, China, India, Indonesia, Malaysia, Vietnam, Singapore, Brazil, Mauritius, Taiwan, the Philippines, Thailand as well as West Africa.

Commercial Production Centers

Commercial scale cultivation of patchouli plant is common in South East Asian countries like Indonesia, Malaysia, China and India.

Growing Practices for Patchouli

Introduction

Patchouli can be grown as a plantation crop as well as an intercrop in plantations. It is a good intercrop in mango orchards, coconut plantations, and other fruit trees such as custard apple. A detailed description of commercial growing practices for patchouli herb is given below:

Site Selection

The first step in growing patchouli is to choose a suitable site where plenty of sunlight is available for healthy growth of the plants.

Patchouli grows well under open field conditions where plants are partially shaded by shade-providing trees. A site free of weeds, water-logging, and soil erosion is the most ideal requirement for growing patchouli on a commercial scale. Soil should be highly fertile and rich in organic matter. Site should also be free of nematode infestation as patchouli plants are susceptible to nematode attack.

Total available area in the site needs to be divided into three major areas – nursery bed, cultivation area and drying area. Nursery bed and drying area may be established in the northern corner of the main cultivation area as maximum shade is available in this location. Main cultivation area is an open field which needs to be well-prepared before planting process.

Cultivars (Varieties)

Patchouli cultivars are often named after their countries of origin. A description of patchouli cultivars, their countries of origin and cultivar characteristics is given below:

Cultivar/Variety	Country of Origin	Description
Java	Indonesia	Indonesian variety that produces more herbage and oil yield but oil quality is inferior to Johore
Singapore	Singapore	This variety produces more herbage and therefore yields more quantity of oil per unit weight as compared to other varieties. However, oil quality is inferior to Johore
Johore	Malaysia	Malaysian variety that produces superior quality oil in terms of chemical composition and odour characteristics. It fetches higher price in the global market
Samarth	India	Released by CIMAP (Center of Medicinal and Aromatic Plants), India

Climate

Patchouli herb grows well up to an altitude of 800-1000 meters above MSL (mean sea level). Ideal rainfall requirement is ranging from 150-300 cm per annum. Patchouli crop prefers an even distribution of rainfall throughout its growth.

Most ideal temperature range is between 25°and 35°C. Ideal atmospheric humidity requirement is about 80-90 percent. In general, patchouli herb prefers a warm and humid climate or tropical to subtropical climate with plenty of sunlight.

Soil

Patchouli herb prefers well-drained, fertile, deep loam to sandy loam soil which is neutral to slightly acidic in nature. Ideal pH is between 5.5 and 6.5. Water logged soils must be avoided because they are susceptible to nematode attack.

Patchouli crop can be grown in alkaline soils also provided that the soil is treated with gypsum@10-15 tons per hectare to modify soil pH.

Light

Patchouli is a shade-loving plant and loves filtered sunlight. Therefore direct sunlight should be avoided. This can be done by planting shade-providing trees on the site before establishing a patchouli plantation.

Patchouli grows well under partial shade. Complete shade is not desirable for its growth.

Propagation

Propagation in patchouli can be done either through rooted stem cuttings or through rooted and hardened tissue culture plants. Generally, rooted stem cuttings are preferred as they are cheaper than tissue cultured planting materials. Opting for tissue cultured planting materials increases initial investment even though tissue cultured plants produce more biomass than that of plants raised through stem cuttings.

Stem cuttings are raised in a well-prepared nursery bed. Tissue culture seedlings are produced in a tissue culture lab.

Preparation of Nursery Bed

Since patchouli is a shade-loving plant, nursery bed needs to be raised in a shaded location. Well-prepared, infestation-free growing media (i.e. fertile soil or soil and organic manure mixture) should be used in the nursery beds. Generally, a nursery bed of 200 square meter size is sufficient to raise planting materials for planting one hectare area.

Preparation of Stem Cuttings

10-15 cm long stem cuttings from healthy branches are prepared to plant in the nursery beds. Each cutting needs to have at least 4-5 plump (healthy) nodes. All leaves are removed except one or two pairs of leaves that are at the top.

Growth stimulators may be applied at the cut ends of the cuttings to improve sprouting percentage. Generally, IBA (indole-3-butyric acid) @1,500–2,000 ppm is used for sprouting enhancement.

Treated stem cuttings should then be planted in nursery beds at a spacing 10 cm × 10 cm. Ideal time for raising a nursery is rainy season. Seed pans or polythene bags containing well-prepared growing media may also be used to raise stem cuttings.

Once cuttings are planted, nursery beds are kept continuously moist by regular watering. Humidity at the root zone is essential for early rooting. Nursery beds are kept moist until the cuttings are ready for transplanting. Under favourable conditions, about 85-90 percent cuttings begin to develop roots within two weeks of planting. Cuttings become ready for transplanting in the field within one to two months.

Transplanting

In case of nursery-raised stem cuttings, 2-month old rooted cuttings are used for transplanting. Rooted cuttings are transplanted in the field in the evenings.

In case of tissue culture seedlings, they need to be adequately hardened before transplanting them in the main field. They can be hardened by gradually exposing them from lab to field by taking appropriate measures. Normally, hardening for a period of one month is sufficient.

Spacing

Recommended spacing for patchouli plants for different commercial growing practices is as given below:

1. 60 cm × 60 cm for rooted stem cuttings for open field cultivation under both drip and conventional irrigation
2. 30 cm x 30 cm for tissue culture seedlings for open field cultivation under both drip and conventional irrigation
3. 45 cm x 45 cm for intercropping system

Planting Time

In case of rooted stem cuttings, the best time for planting them in the main field is onset of monsoon; i.e. June to September. In case of tissue culture seedlings, any time except very hot and heavy rainy seasons is ideal.

Planting Depth

Planting materials (either rooted stem cuttings or hardened tissue culture seedlings) are planted directly on ridges that are prepared in the main field at a planting depth of approximately one centimeter.

Field Preparation

Field or main site of planting should be well-prepared well ahead of planting process. This may be done by 3-4 ploughings followed by leveling of the land. Top soil should be mixed with 10 to 20 tonnes of FYM (farm yard manure) or compost during last ploughing to enhance top soil fertility. The field, then, is laid out into ridges or beds.

Ridges or raised beds of convenient size are prepared. Generally, ridges (beds) of having 20-25 cm height and 15-20 cm width with a spacing of 60 cm between the rows are recommended. However, a grower should always follow his/her discretion while preparing land and ridges, based on his/her convenience, land and soil characteristics, availability of natural resources etc. The ridges or beds need to be irrigated a day before the transplanting process.

Since nematode attack is a serious problem in growing patchouli in certain regions, application of neem cake or pongamia cake or any other locally available organic nematicide @ 0.5 tons per acre is recommended at the time of land preparation. Another practice to prevent nematode infestation in the soils is to grow nematode-repellent crops such as periwinkle along the borders of the ridges.

In regions where soil fertility is poor, basal application of 8, 20 and 20 kg of N, P and K per acre is also recommended during field preparation.

Fertilizer and Manure Application

Patchouli is a heavy feeder and hence fertilizer requirement is heavy. Fertilizer and manure application needs to be scheduled after proper soil testing procedures to determine soil fertility status. Patchouli is a heavy feeder of nitrogenous fertilizer and hence nitrogenous fertilizers needs to be applied in liberal quantities.

For 3-4 harvests a year, the plant has to be fed with 150:100:100 Kg of NPK (N for Nitrogen, P for Phosphorus and K for Potash) per hectare. Full P and K and 1/4 N is applied as a basal dose along with organic manures at the time of land preparation. Rest of the Nitrogen is applied in 3 equal splits after each harvest if three harvests are taken in a year. Sometimes, foliar spraying of urea (0.2%) is recommended as an additional dose.

Foliar application of micronutrients is recommended in case the plantation is susceptible to chlorosis and browning.

In soils deficient in trace elements such as zinc and manganese, application of zinc sulphate@20 kg/acre and manganese sulphate@10 kg/acre is recommended.

In advanced, intensive growing systems by using drip irrigation or other advanced technologies such as hydroponics, higher doses (generally 1.5 times) of manures and fertilizers are recommended as bio mass production is more under such intensive growing systems.

Irrigation

Patchouli is a moisture-loving plant and irrigation is an important practice in growing patchouli plants. Field beds (ridges) are irrigated one day before transplanting process. Thereafter, a light irrigation is done soon after transplanting process.

Patchouli plants need shade and sufficient moisture during initial stages of its growth to get established in the field. Generally, daily watering is recommended for the first week after transplanting and thereafter watering may be done every alternate day for the next two weeks or till the time plants are established.

While watering patchouli plants, care needs to be taken to avoid overwatering as patchouli plants are susceptible to waterlogging. Waterlogged soils may invite nematode infestation also.

Once plants are established in the soil, irrigation may be scheduled depending upon water holding capacity of the soil and climatic conditions.

Patchouli can be grown either by using conventional irrigation methods such as flood irrigation or by using drip irrigation. Frequency of flood irrigation may be scheduled as twice a week and that of drip irrigation as 30 minutes per day. In drip irrigation, fertilizers may be applied through irrigated water (i.e. fertigation may be practiced).

During summers, plants need more water and therefore frequency of irrigation needs to be doubled. During monsoon season, irrigation is not required.

Nipping or Pinching

Once plants get established in the field, nipping of apical buds is done at monthly intervals in order to promote branching and vigorous lateral growth of the plant.

Disease Management

Patchouli is susceptible to leaf blight disease, a fungal infection caused by Cercospora species and rhizoctonia wilt, a fungal infection caused by Rhizoctonia species. In leaf blight disease, main symptoms include appearance of brown spots near leaf margins of almost one year old plants. The affected leaves dry up gradually. In rhizoctonia wilt, the affected plants wilt and gradually die. Application of recommended fungicides is the best control measure that can be adopted by a grower in both cases.

Pest Management

Patchouli crop is susceptible to root-knot nematode (*Meloidogyne incognita*) infestation. Main symptoms include wilting of the plants, stunted plant growth and presence of galls in the roots. Pre-planting treatment of planting materials with organic nematicides such as neem cake or pongamia cake is recommended as a control measure. Using a healthy mother stock to raise cuttings under nematode-free conditions also eliminates the chances of nematode infestation. Other nematode control measures include sterilization of nursery beds, growing periwinkle, a nematode-repellent crop as an intercrop, and use of recommended nematicides. In organic cultivation practices, biofertilizer containing *Trichoderma spp.* may be applied in large quantities at the time of land preparation to counteract the effect of nematode attack.

Apart from nematodes, aphids and caterpillars may also become serious problems in a patchouli plantation. These problems can be controlled by using any of the recommended pesticides.

Weed Management

During initial two to three months of plant growth, the field needs to be kept weed-free by manual weeding. Hoeing is also recommended during initial stages of plant growth. During hoeing operation, care needs to be taken not to disturb the root system of growing plants. In an established plantation, weeding and hoeing is recommended after each harvest.

Growth Duration

It may take about 6 to 9 months after planting for a patchouli plantation to come to maturity for harvesting.

Harvesting

When the lower leaves of the plant turn brown in colour, it is time to harvest the leaves. Since the highest quality oil is obtained from new emerging shoots, these top pairs of leaves are harvested from the top. Normally, leaves and tops are harvested at 6 to 8 nodes or 10–25 cm below the apex point. If harvesting is done further below this point, subsequent yields will be less. Harvesting is done during late morning hours by using a sharp knife.

In case of rooted stem cuttings, first harvest is possible after six months of planting; that is, when the plants are about one meter tall. In case of tissue culture plants, plant maturity is quick and first harvest is done three months after planting. Subsequent harvests in both cases may be possible at an interval of 3-4 months, depending upon the soil fertility status, prevailing climate and cultural management practices.

Economic Life

A patchouli plantation yields quality produce up to 3 years.

Yield

Under flood irrigation, approximately 10 tonnes of fresh leaves is obtained from one acre area per annum in 3 harvests. That is, approximately 3 to 3.5 tons of fresh leaves per harvest are obtained.

Under drip system of irrigation, approximately 20 tons of fresh leaves are obtained from one acre area per annum in 3 harvests. That is, approximately 5 to 7 tons of fresh leaves are obtained per harvest.

While growing patchouli as an intercrop, a yield of about 2 tons of biomass is obtained per acre per year.

Drying Process

Freshly harvested leaves are dried in thin layers in shade. The leaf layers are turned frequently to prevent rapid fermentation and thus to ensure proper drying. For higher oil recovery and good quality of oil, moisture content of herbage should be between 8-10 %. Drying normally requires 3-6 days. Properly dried leaves develop characteristic patchouli aroma, which is less noticeable in fresh leaves. Shade-dried leaves are stored in well packed gunny bags for at least 3 months for ageing process. Ageing improves the odour. Mature stalks are removed before distillation as these stalks contain no oil. Oil percent in dried leaves is 3 to 3.5%.

A yield of 4 tons of fresh leaves per acre per year when dried under shade for 3-6 days yields 1- 1.25 tons of dry leaves per acre per year.

Extraction of Patchouli Oil by Steam Distillation

Dried leaves are steam distilled for better oil recovery. Maintaining recommended steam pressure inside the distilling unit is essential for maximum oil recovery. An interchange of high and low pressures i.e. 1.4 to 3.5 kg/m2 produces better quality oil.

The recovery of oil from the shade dried herb varies between 3 - 3.5 per cent. The duration of distillation is 6 to 8 hours for complete recovery of the oil. Properly dried leaves produce good oil yield and better quality of oil.

Oil Yield

One ton of dried leaves (from one acre area) yield up to 30 to 40 litres of oil.

Oil Purification

Freshly extracted patchouli oil contains moisture and impurities which need to be removed by filtration method. Moisture present in the oil deteriorates oil quality; hence moisture needs to be removed from the oil by adding anhydrous sodium sulphate @ 20–30g/liter and keeping the distillate mixture for 4–5 hours. After the waiting period, oil is filtered to remove all impurities.

Oil Characteristics

Good quality patchouli oil is a yellowish to reddish brown clear liquid with a rich, sweet, dry leafy (campharaceous) herbal woody aroma and variable viscosity. It gets better in quality and develops its characteristic odour with aging. Two important components of patchouli oil are patchoulol and norpatchoulenol. Among these, major active ingredient is Patchoulol (up to 50 percent) which imparts the oil its characteristic flavor and aroma. Patchouli oil has a very characteristic aroma and blends well with other essential oils such as sandalwood oil, vetiver oil, clove oil and geranium oil.

Storage and Packing of Oil

It is to be ensured that oil does not contain any water before storage. The oil is stored in glass bottles or drums made up of steel or aluminium depending upon the quantity of oil to be stored. The containers are filled up to the brim, tightly capped and stored in a cool, dry and dark place.

Patchouli oil when comes in contact with the outside air may oxidise and as a result the quality of oil may deteriorate. So care should be taken to store the oils in air tight bottles.

Economic Importance of Patchouli Oil

Commercial Uses of Patchouli Oil

Patchouli oil in Aromatherapy
Patchouli oil is highly therapeutic and is used in aromatherapy for curing stress-related lifestyle problems. It calms nerves and provides general relaxation.

Patchouli oil in Perfumery
Patchouli is used widely in perfumery industry as an ingredient for manufacturing some well known oriental perfumes. Since the odour of patchouli oil is strong and herbaceous with a spicy fragrance and it blends well with other aroma oils of sandalwood, geranium, vetiver, clove etc, it is highly favored as an ingredient in perfume manufacturing, especially for making heavy perfumes. Since there is no substitute for patchouli oil and it cannot be synthetically replaced, patchouli oil has a unique place in perfumery industry.

Patchouli oil as an Insect-Repellent
Patchouli oil may be used as an all-purpose insect repellent.

Patchouli oil as an Ingredient in Incense
Patchouli oil, sometimes in combination with sandal wood oil, is used as an important ingredient in making some famous brands of East Asian incense sticks.

Patchouli oil as an Ingredient in Breath-Freshener
Patchouli oil is an important ingredient in some brands of breath fresheners. It is often blended with anise and clove to get desired results.

Patchouli oil as an Ingredient in Cosmetics and Toiletries
Since patchouli oil has a strong pleasant odour, it is used as a fragrance in cosmetics, soaps, detergents, after-shave lotions and many other toiletry products.

Patchouli oil as an Ingredient in Cigarettes
In combination with sandal wood oil, patchouli oil is used in blending of tobacco.

Food Uses of Patchouli Oil

In very low concentration (2 ppm), patchouli oil is used as a food flavouring agent. It is used to flavour foods, beverages, candies, baked products, meat preparations, desserts and chewing gum.

Therapeutic Uses of Patchouli Oil

Patchouli oil has several medicinal properties. Major among them are,

1. Antifungal: Patchouli oil is a natural remedy for fungal infections such as athlete's foot
2. Anti-inflammatory: Patchouli oil reduces inflammations and swellings when applied externally
3. Antiseptic: Patchouli oil destroys skin infections
4. Aphrodisiac: Patchouli oil promotes reproductive and sexual health
5. Antiviral: Patchouli oil kills viruses
6. Anxiolytic or Anti-depressant: Patchouli oil is used as an anti-panic or anti-anxiety agent. Patchouli oil heals problems due to mental agitation such as tension, insomnia and anxiety
7. Rubefacient: Patchouli oil improves skin blood circulation. It is a well known tissue regenerator which stimulates growth of new skin cells
8. Hair treatment: Patchouli oil treats dandruff

Effect of Patchouli Oil on Physical Health

Patchouli oil is used as health-stimulating agent to alleviate pains; destroy fungal, viral, and bacterial skin infections; improve blood circulation throughout the body; encourage sweating to cure fevers; strengthen digestive system; and for healthy skin growth.

Effect of Patchouli Oil on Mental and Emotional Health

Patchouli oil is known for their calming, relaxing and stress-relieving properties. The oil may also be used as a natural remedy for many skin problems such as open pores, acne and pimples, eczema, cracked and bruised skin, and wrinkles.

Therapeutic Uses of Patchouli Fresh Leaves
In Chinese traditional medicines, a decoction prepared from fresh leaves of patchouli is used as a home remedy to treat nausea, vomiting, cold, headaches and diarrhea. In some countries, a paste made from fresh leaves is used to heal burn.

Uses of Patchouli Dried Leaves
Dried leaves of patchouli are used for pot-pourri.

Market Potential of Patchouli Crop

Since patchouli oil is an essential ingredient in several perfumeries, cosmetic and toiletries products, and pharmaceuticals, it has a growing demand in global market. Currently Indonesia is a leading producer and exporter followed by Malaysia, China and Brazil. European Union and the United States of America are major export markets for patchouli oil.

References

CIGR Journal. (2014, May 5). Retrieved from
http://www.cigrjournal.org/index.php/Ejounral/article/vie
wFile/2289/1746

National Bank for Agricultural and Rural Development, India.
(2014, May 5). Retrieved from
https://www.nabard.org/english/medical_patchuoli.aspx

National Horticulture Board, India. (2014, May 5). Retrieved
from http://nhb.gov.in/Horticulture

www.ingramcontent.com/pod-product-compliance
Lightning Source LLC
Chambersburg PA
CBHW051830170526
45167CB00005B/2222